MW00747994

This book has been published in cooperation with Evans Publishing Group.

Published in the United States by
Amicus
P.O. Box 1329, Mankato, Minnesota 56002

Printed in China by New Era Printing Co.Ltd

Library of Congress Cataloging-in-Publication Data

Graham, Ian, 1953-
Technology Careers / by Ian Graham.
p. cm. -- (In the workplace)
Includes index.
Summary: "Describes aspects of working in the field of information and communication technology. Includes information on computers, the internet, and communication and broadcasting"--Provided by publisher.
ISBN 978-1-60753-095-4 (library binding)
1. Information technology. 2. Telecommunication. I. Title.
T58.5.G722 2011
004--dc22

2009045300

Editor and picture researcher: Patience Coster
Designer: Guy Callaby

We are grateful to the following for permission to reproduce photographs: Alamy 8 (Maximilian Stock Ltd/Phototake), 10 (PhotoStock-Israel), 14 (67photo), 16 (Moodboard), 17 (Mira), 19 (Archimage), 20 (Dave Jepson), 21 (Peter Treanor), 22 (Cris Haigh), 30 (Peter Dazeley), 33 (Lloyd Sutton), 34 (David R. Frazier Photolibrary, Inc.), 36 (Dennis MacDonald), 37 (Tony West), 41 (AfriPics.com), 42 (Dennis MacDonald); Corbis 11 (TWPhoto), 12 (Alexandra Beier/Reuters), 13 (Virgo Productions), 15 (Louie Psihoyos), 23 (Krista Kennell/Electronic Arts), 25 (TWPhoto), 27 (Ryoichi Utsumi/ Amanaimages), 29 (Paulo Fridman), 31 (Jennifer Taylor), 35 (Catherine Karnow), 38 (Helen King), 43 (Marco Cristofori); Getty Images 6 (Digital Property), 7 (Lester Lefkowitz), 9 (Andy Sacks), 39 (ColorBlind Images); iStockphoto 26, 28, 32, 40. SPL cover and 18 (Thierry Berrod, Mona Lisa Production/Science Photo Library), 24 (Gustoimages/Science Photo Library).

05 10
PO1568

9 8 7 6 5 4 3 2 1

Technology Careers

IAN GRAHAM

amicus
mankato, minnesota

Contents

Working in Technology

The technology field is a wide-ranging industry. Information technology (IT) covers technology for storing, retrieving, manipulating, and transmitting data. Communication technology includes computers, the Internet, telephones, wireless communications, and broadcasting. Together these fields are called ICT, or Information and Communication Technology. The technology ranges from small devices in household appliances and vehicles to office computer systems, supercomputers, and international telecommunications networks. All sorts of industries and businesses use ICT and they all need people to design, install, maintain, repair, and upgrade their systems. Some companies have their own ICT staff. Others contract with service companies to meet their ICT needs.

SHOWCASING SKILLS

Some technology jobs can be entered from high school or technical school as an apprentice, but many others, such as engineering jobs, usually require a bachelor's degree. Work experience can improve your chances of success in finding your first job.

Good computer skills are essential for most careers in ICT.

Training courses and university degrees that include some time spent working in industry are particularly valuable. Volunteer work involving computers, electronics, web sites or communications can help too. If possible, make a CD, DVD, or web site to showcase your work and skills. Some large companies and other organizations offer apprenticeships or on-the-job training leading to ICT careers.

ICT professionals are trained to work with a wide variety of electronic equipment and systems.

A GLOBAL INDUSTRY

There are a huge variety of technology careers, including computer programmer, game developer, web designer, systems analyst, communications engineer, and information scientist. Some of these careers are open to graduates in any subject, but many of them require degrees in an ICT-related subject such as math, physics, electronics engineering, or computer science. It is important to check the entry requirements for your chosen career. It isn't essential to know a foreign language, but languages are becoming increasingly attractive to employers, especially large corporations, because ICT is a truly global industry. Suppliers, customers, engineers, and managers in different countries need to be able to talk to one another, so a foreign language could be a useful addition to the qualifications you can offer an employer. ICT professionals with some experience behind them can move into more senior positions or start their own businesses. There are also opportunities for telecommuting. This allows ICT workers such as programmers, who spend a lot of time working at a computer, to log in to their employer's network from anywhere and work without having to go to the office.

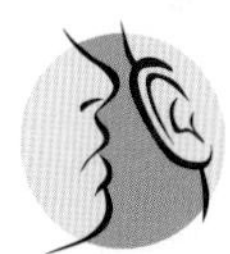

HANDY HINT

Training and learning don't end when you get a job. ICT changes all the time, because new technology is constantly being developed. ICT professionals keep learning throughout their careers. Training courses can lead to certification, showing that an engineer or consultant is qualified to a certain standard. A wide range of professional organizations and technology suppliers offer certification.

Chips and Circuits

Electronic circuits are used in a huge variety of products and equipment. They're in cars, aircraft, ships, kitchen appliances, TVs, telephones, digital cameras, computers, spacecraft, and all sorts of industrial equipment.

ELECTRONICS ENGINEER

Electronics engineers are involved at every stage of developing electronic products and systems. They design new electronic circuits, build them, test them, and take them right up to manufacture. In industry, electronics engineers also install electronic systems and maintain them.

TO WORK AS AN ELECTRONICS ENGINEER, YOU WILL NEED

- *good problem-solving skills*
- *the ability to work on a team*
- *attention to detail*
- *science and math skills*

SOLVING PROBLEMS

The work to develop a new product begins with a brief–a description of what's needed. Electronics engineers work closely with clients to learn what their requirements are. The electronics engineer chooses the electronic components and designs the circuit board that connects them together. The circuit may be designed with the help of a Computer Aided Design (CAD) package. The next step is to build prototypes and test them. Any problems or glitches that show up at this stage have to be solved, so electronics engineers need good problem-solving skills. When the client is happy with the final working prototype, the product can be manufactured. Products rarely stay the same for long. They are soon modified or improved as the client's needs change. If any alteration to the electronics is needed, electronics engineers do this work.

Electronics engineers design and build electronic circuits.

A HIGHLY VARIED CAREER

Working in electronics can be highly varied. There's a lot more to it than building electronic circuits. Only part of an engineer's time is taken up with building circuits and testing them. Some of the work may involve visiting customers to provide support and maintenance. There may be opportunities for foreign travel too, to meet other engineers, managers, and partners in an organization to discuss projects. Good communication skills, both spoken and in writing, are important because the work involves presenting the results of projects, writing reports, and explaining new products to managers and engineers.

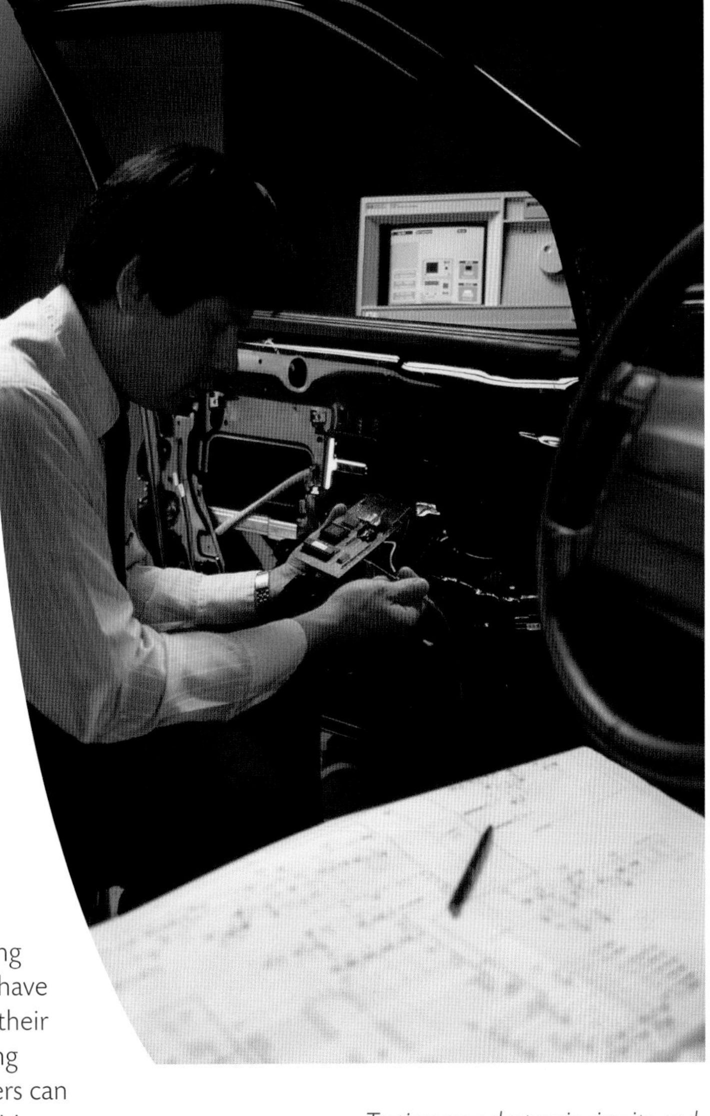

Testing new electronic circuits and products is an important part of an electronic engineer's work.

KEEPING UP-TO-DATE

A degree in math, physics, electronic engineering, or a similar subject is usually needed for a career in electronics. The technology is changing all the time, so electronics engineers have to keep studying and training during their career to stay up-to-date. After gaining some experience, electronics engineers can hope to move on to more senior positions managing other staff and working on bigger projects. Large organizations may offer training courses, in project management for example, and temporary placements in different parts of the organization, such as marketing or finance. These courses and placements give a broader understanding of the whole business.

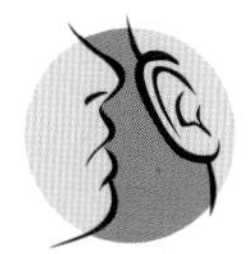

HANDY HINT

Good color vision is essential because some electronic components and wiring are color-coded.

Carol: Electronics Engineer

"I was always fascinated by electronic gadgets and I was good at math and science at school, so it seemed natural to study electronics engineering at the university. When I graduated, I was recruited by a large consumer electronics manufacturer. I joined the company's graduate trainee program. My first job was as part of a project team developing wireless products. I was surprised by the amount of responsibility I was given. I was working in the lab right away, testing prototypes of circuits and products the team had already produced and working on improvements to them. I really enjoyed the problem-solving part of the work. That's probably why I became an engineer.

"The company has research labs in Italy and I have to go there to present the results of my work to other engineers. It gives me a real sense of achievement. The work is quite varied. As well as the practical work, I have to write reports, attend planning meetings, and research products made by other manufacturers. Sometimes I'm working on my own, but more often I'm working alongside other members of the team.

Problem-solving is part of an electronics engineer's work.

"One advantage of working for a big company is that they offer lots of training courses that will help to advance my career. I'm taking one in project management that will help me to run my own project team in the future."

COMPUTER HARDWARE ENGINEER

Computer hardware is the chips, circuit boards, screens, printers, hard drives, and other equipment that make up a computer system. It's all the "stuff" you can pick up and hold in your hands. Computer hardware engineers are electronics engineers who specialize in working with computer equipment. They are involved in the design, development, installation, and maintenance of computer hardware. Advances in the design of computers, from laptops to supercomputers, are the result of the work of computer hardware engineers.

DESIGN AND DEVELOPMENT

Nearly half of computer hardware engineers work for manufacturers of computers and other computerized equipment. Some of them design new microprocessors and other computer chips. Others concentrate on particular parts of computers, such as hard drives or keyboards, or they may produce whole computer systems. They work in offices and workshops, building and testing the prototypes of new computer equipment before full-scale manufacturing begins. The work begins with a specification, a description of what the component or computer will have to do. This is used to produce detailed circuit diagrams. Hardware engineers use the diagrams to build a prototype, which they test to make sure that it does exactly what the specification asked for. Lots of computers are "embedded" in (built into) other machines, including cars, ships, planes, communications equipment, entertainment systems, and office machines. Computer hardware engineers develop these embedded computers, too.

Engineers work in teams to develop new products.

Forensic computer analysts examine computer equipment seized during an investigation for evidence of criminal activity.

ENGINEERING SPECIALISTS

Computer engineers can specialize in lots of different ways. Design engineers concentrate on the design of new computer equipment. Product development engineers are involved with turning designs into products that people can use. Test engineers test computer systems to check their performance and look for glitches. Forensic computer analysts help the police and other law enforcement agencies find evidence in crimes that involve computers. These are just a few of the many types of specialized work that computer engineers can do.

FINDING A JOB

Computer hardware engineers are employed by a great variety of businesses and industries, including computer manufacturing, telecommunications, transportation, security, and robotics. A degree in a relevant subject such as electronics engineering, computer science, or computer engineering is usually needed. A master's degree or a doctorate may be necessary for more senior or specialized jobs. Students can prepare by taking courses in subjects such as IT, computers, math, physics, and electronics.

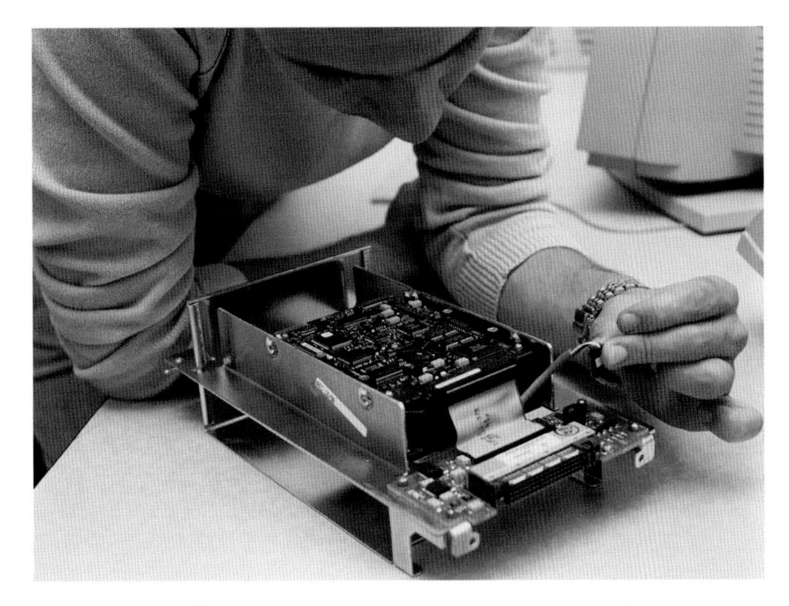

A computer hardware engineer builds, installs, and troubleshoots computer equipment.

MAIN TASKS: COMPUTER ENGINEER

- *designing computer hardware*
- *installing computers*
- *testing and repairing computers*

TO BECOME A COMPUTER HARDWARE ENGINEER, YOU WILL NEED

- *technical knowledge of electronics and computers*
- *the ability to work alone or on a team*
- *attention to detail*
- *good communication skills*
- *good color vision*

COMPUTER OPERATOR

Computer operators control computers, mainly large computers such as mainframes and supercomputers. In the early days of computing, the ordinary workers in an organization did not operate computers themselves. Instead, the computers were run by teams of specially trained operators. As computers became faster, with much bigger memories and more advanced software, more of the operators' work was automated and ordinary workers had access to computers. Today, there is a computer on almost every desk. However, there is still some need for computer operators, especially in big organizations that have to process large amounts of information. Organizations such as government departments, health services, scientific research institutes, law enforcement agencies, and banks all use large computers and need operators to look after them.

MONITORING PERFORMANCE

A computer operator's work depends on the size of the organization and the type of computer system. It may include starting up and closing down the computer system, unless it runs 24 hours a day. The operator runs programs and loads data into the computer, makes sure that each job is successfully completed, and prints out the results. The work also involves checking that the computer system is running properly, monitoring its performance, and ensuring that all the data is backed up securely. If the system runs 24 hours a day, shift work may be necessary to provide a 24-hour service.

ROUTINE WORK

A computer operator's work can be calm and routine for long periods, but when a system crashes or a problem occurs, the operator has to identify the problem quickly and solve it or pass it on to support staff without delay. If a scanner, printer, or another piece of equipment breaks down, the operator has to switch the work to alternative equipment so that service is not interrupted. Jobs are likely to come in from departments in different parts of an organization. Operators prioritize the work and schedule it so that it all gets done on time. They usually have to log (keep records of) when jobs are received, when they are run, and when results are delivered, so accurate record keeping is important.

TO BECOME A COMPUTER OPERATOR, YOU WILL NEED

- *good computer skills*
- *a willingness to do shift work*
- *a working knowledge of operating systems and hardware*
- *attention to detail*

Computer operators run the computer systems of large organizations.

Supercomputers, the fastest and most powerful computers of all, are operated by dedicated professionals.

ON-THE-JOB TRAINING

Entry to a computer operator's job usually requires a high school diploma. Training is normally on-the-job, because of the different hardware, software, and work practices used by different organizations. After a few months of training, a computer operator then usually works unsupervised. A reasonable level of fitness and strength is needed, because part of the operator's job is to keep printers supplied with ink and paper, and boxes of printer paper can weigh more than 45 lbs. (20 kg).

TECHNICAL SUPPORT SPECIALIST

Technical support specialists diagnose problems in computer systems and repair them, or advise computer users on how to solve them. When any of an organization's computer systems develops a problem, the technical support specialist is the person called in to help. Technical support specialists may be known by other names, including technical support technicians, support engineers, or tech support. Support technicians who are based at a desk, where they answer requests for help and advice by phone and e-mail, are called help-desk technicians. Technical support is essential if a business is to use its computer resources efficiently.

WHERE WILL I BE?

With experience, computer operators can move up to more senior operator positions, supervising other staff, or to other roles in IT network maintenance and technical support.

Technical support specialists diagnose the causes of computer crashes.

ALL-AROUND KNOWLEDGE

Support technicians may visit in-house staff or go out to clients and carry out minor repairs themselves. They need a good all-around working knowledge of computer hardware, software, and operating systems, because problems can arise in any part of a computer system. Technicians find out the nature of a problem by asking computer users questions and analyzing their answers. They test the computer equipment themselves too, sometimes using diagnostic software to identify the source of a problem. They keep records of the technical glitches they deal with and how they were solved for future reference. Good communication skills are essential because support technicians may have to deal with staff who are feeling quite stressed by the breakdown of their computer equipment. Tact, patience, and calm professionalism are important qualities in these situations.

ON-SITE SUPPORT

Technical problems can happen at any time, so technical support specialists may have to work shifts or on weekends to ensure that systems are covered all the time. Depending on the size and nature of the business, support technicians may work in one location, or they may go out to visit clients and provide on-site support. Diagnostic and repair work is usually urgent and support staff are often under pressure to bring computer systems back online as fast as possible. Successfully diagnosing a complex problem and resolving it quickly can be very satisfying.

MAIN TASKS: TECHNICAL SUPPORT SPECIALIST

- ***providing technical assistance***
- ***analyzing computer problems***
- ***using diagnostic software***
- ***answering technical questions***
- ***repairing equipment***
- ***installing equipment***
- ***upgrading equipment***

DIPLOMAS AND CERTIFICATES

There are several ways to enter technical support work. It may be possible to enter straight after high school as an entry-level technician. Certifications that demonstrate an interest in and knowledge of computer systems will improve an applicant's prospects. Larger businesses may offer IT, engineering, or electronics apprenticeships that lead to work in technical support. However, some employers prefer applicants to have an associate or bachelor's degree in computer science or a related field.

WHERE WILL I BE?

A technical support specialist may be promoted to team leader, supervising a support team. The skills a technical support specialist acquires can be applied to other IT positions, so a technical support officer could move into other areas of IT such as training, product development, or network engineering.

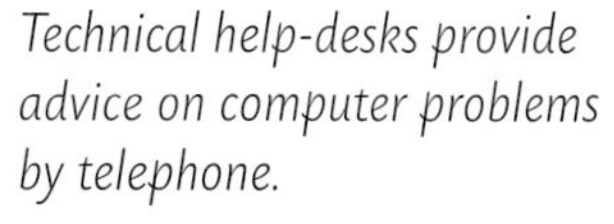

Technical help-desks provide advice on computer problems by telephone.

Working with Software

Computer hardware is nothing without software, the programs that make computers work. Computer programming is one of the most popular careers in information technology. About a third of all IT professionals are programmers.

COMPUTER PROGRAMMER

A computer programmer writes the programs that enable computers to keep track of a company's finances, play games, control a power plant, guide an airliner, connect telephone calls, or any of the thousands of other things that computers do. A computer programmer may also be known as a software engineer or software developer.

MAIN TASKS:
COMPUTER PROGRAMMER

- *designing and writing new software*
- *testing new software for bugs*
- *modifying existing software*
- *developing documentation*

Programmers work closely with technical staff and managers to produce software.

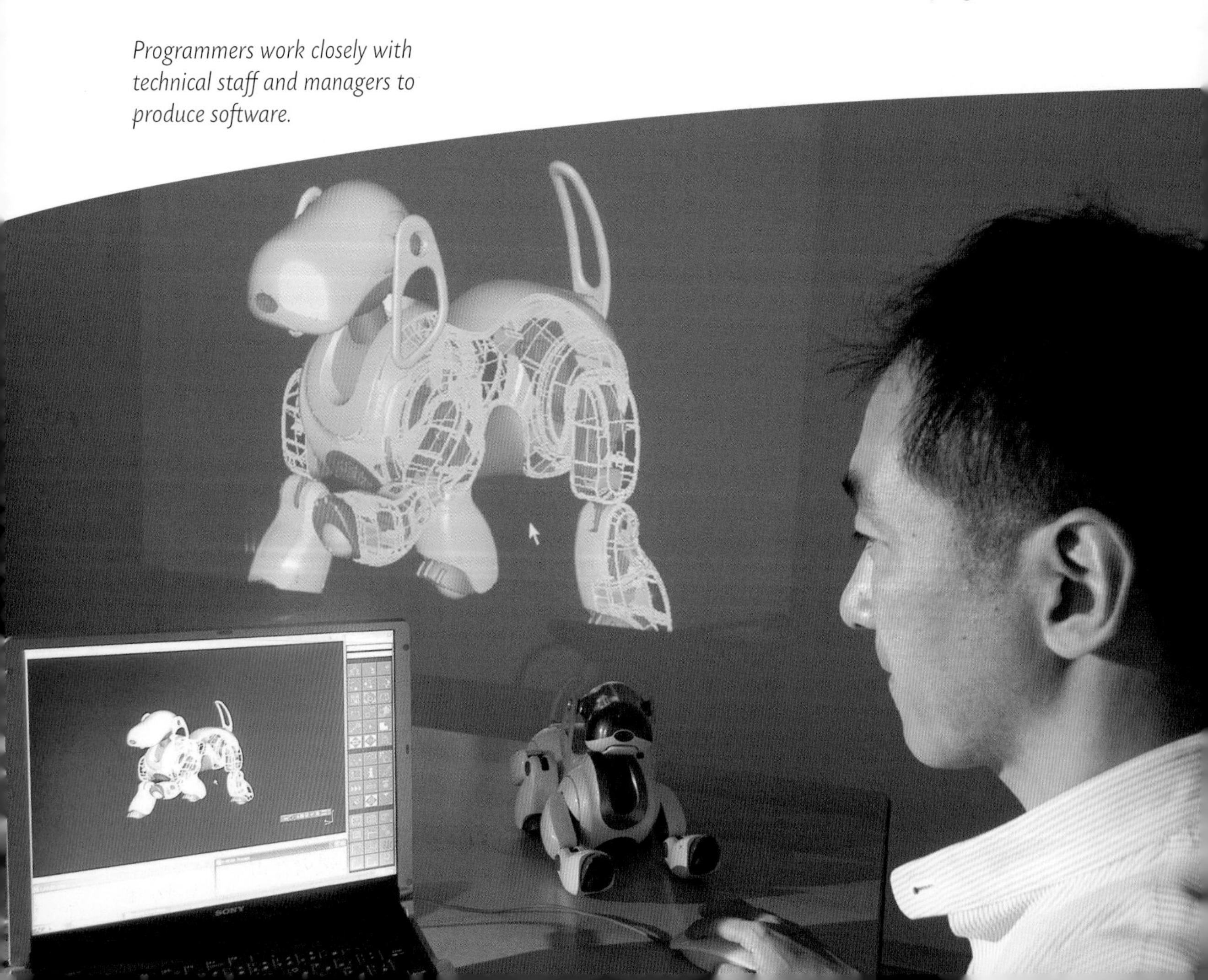

Programmers write the software that monitors and controls large complexes such as power plants and operates the communications systems used by organizations such as ambulance services.

WHERE WILL I BE?

With a few years experience, a computer programmer can become a senior programmer or lead programmer with more responsibility. A few years later, a senior programmer can move up to become a project manager or a software architect in charge of major projects. Some programmers go on to start their own businesses providing software services as consultants.

ANALYZING PROBLEMS

Programmers write software to meet a particular need or to solve a specific problem. Before writing a piece of software, a programmer has to understand exactly what it will have to do. It might be needed to control machinery in manufacturing, operate a new music player, analyze customer information for a business, or monitor an alarm system. The programmer studies the problem and works out a solution. The solution is broken down into a series of simple instructions called an algorithm. Then the programmer writes the software to carry out the instructions. For some projects, a programmer might use off-the-shelf software packages, perhaps adapting them if necessary. On other projects, a programmer may write all of the software from scratch. A large project may involve a team of programmers, who divide up the work among them. Programmers attend regular meetings to discuss progress and ensure that the project is kept on schedule. When the software has been written, it is not installed and run for real until the programmers have tested it thoroughly and checked it for glitches called bugs.

TO BECOME A COMPUTER PROGRAMMER, YOU WILL NEED

- ***an analytical mind***
- ***good problem-solving skills***
- ***the ability to work on a team***

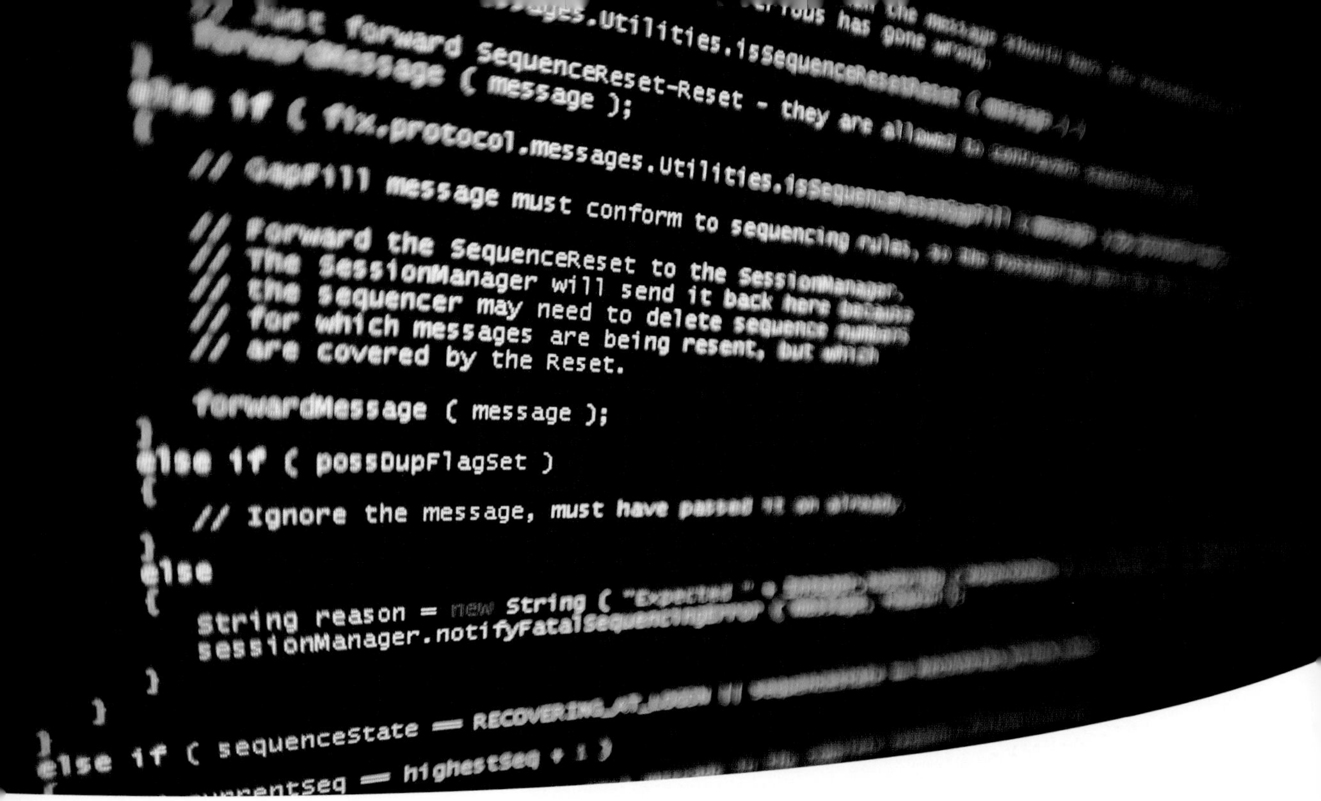

Software engineers spend long hours at a computer writing program code and reports.

LEARNING LANGUAGES

Computer software may be written in one of several programming languages, so programmers need to have a good working knowledge of commonly used languages such as Java, C++, C#, and Visual Basic, and also the main operating systems, such as Windows, Mac OS, Unix, and Linux. When a programmer has written software, the next job is to make a presentation to the people who will use it. The users might be machine operators in industry, office workers, or sales staff. A programmer must be able to communicate with them clearly in nontechnical language. The technical support staff who will have to maintain the software and troubleshoot any problems that arise are trained on it too. Finally, the software is put into operation.

PROGRAMMING SPECIALTIES

Computer programmers often specialize in a particular type of programming. Applications programmers write programs to do a specific job. Systems programmers concentrate more on background software such as operating systems. Multimedia programming is a popular specialization. Programmers write software that brings together text, sound, 2D/3D graphics, photographs, video, and animations in the same package. They use programming languages such as ActionScript, C++, and Visual Basic. They need a combination of creative design and computing skills to do this kind of work. Multimedia specialists, also called multimedia programmers, are employed by multimedia publishers, television and film production companies, web design companies, and software producers.

Terry: Software Engineer

"While I was earning a degree in computer science, I discovered that I was more interested in software than hardware. After college, I went to work for an IT consultant firm supplying IT services to businesses, mainly in the retail and manufacturing sectors. My work involves writing new software and updating existing software for a variety of clients.

"I spend a typical day writing code and testing it. I work regular hours most of the time, although sometimes I have to put in a few extra hours to make sure that deadlines are met. Each job begins with a client conference where we are briefed on what is required. I enjoy the client conferences because I meet a great variety of people and the meetings get me away from my computer for a while. You have to analyze problems and think logically to get the job done. Although the bulk of the work involves writing program code, every job is different and presents its own challenges, so the work is never dull.

"New software products come along all the time and it's important to keep up-to-date, so I take courses to make sure I'm up to speed with the latest developments."

Software engineers test new programs to make sure they run properly.

Game developers make the games they write bright, colorful, and exciting, so that people of all ages will want to play them.

PROJECT TEAMS

Multimedia specialists may work on their own, especially on small jobs. Larger projects are usually handled by a team of specialists with different skills, including writers, artists, animators, filmmakers, programmers, and 3D modellers. The finished product is usually designed to work on a particular media platform, such as DVD, the Web, interactive television, game consoles, or cell phones. A wide variety of degrees can lead to a career in multimedia programming. A degree in math, physics, computer science, software engineering, graphic design, or animation is particularly useful. Another route is to start work as a general programmer or software engineer and then, after gaining some experience, to specialize in multimedia programming.

GAME DEVELOPER

The worldwide market for video and computer games is huge. It's worth more than $40 billion and it continues to grow year after year. The games are created by game developers. They produce games for computers, game consoles, the Web, interactive television, cell phones, and arcades. Their work combines visual creativity with technical ability.

WORKING TOGETHER

Games are rarely produced by just one or two people. They are usually the work of dedicated teams of professionals working for game production companies. Game developers can work as designers, artists, animators, programmers, quality assurance testers, producers, or project managers. They all work together to create a game. Designers decide what a game will look like and how it will play. Artists and animators create the characters and backgrounds. Game programmers spend their working days at computers writing the program code that makes a game work. Quality assurance testers check that the game works. Producers, or project managers, supervise the work and ensure that deadlines are met. From the initial idea to the finished product, a complex game can take a year or more to produce.

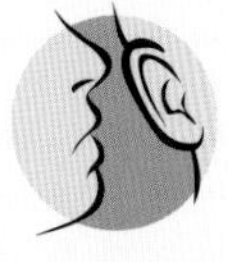

HANDY HINT

An employer will want to see examples of your programming work, so make a demonstration CD or DVD, or set up a web site to showcase what you can do.

ACTION . . . CUT!

The video game industry is moving closer to the movie business. Dozens of feature films based on computer games have already been made and more are in production. The reverse is true too. Successful films have inspired lots of games and produced new career opportunities for people with artistic or digital skills and qualifications that are applicable to both movies and games. In the future, game producers may make more use of scriptwriters and cinematographers—careers normally associated with the movie business. And more movie producers may make use of digital image processing techniques, some of which were created or improved by game developers.

New computer games are sometimes launched at a big show with actors dressed as characters from the game.

GAME-PLAYING BY DEGREE

Game developers come from a wide range of backgrounds, with degrees in math, physics, graphic design, computer science, and computer programming. A degree or certification that includes programming is attractive to employers. It is also now possible to study for a degree in game design, game programming, or game development at some colleges and universities. Alternatively, an undergraduate degree without any programming content can be followed up with a master's degree in game programming.

TO BECOME A GAME DEVELOPER, YOU WILL NEED

- *creativity*
- *computer skills*
- *ability to work with a team*
- *enthusiasm for games*

WHERE WILL I BE?

Promotion can be rapid in the game industry. Entry level game developers can progress quickly to senior developer and lead developer positions. Some game developers also change from one specialization to another—from programmer to producer, for example.

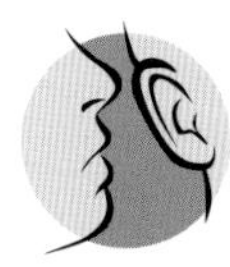

HANDY HINT

Research the games a company has produced, and play them extensively before applying for a job.

Game programs are written to run on popular game consoles.

Game programmers create games using off-the-shelf graphics and animation packages together with software they write themselves.

Andrew: Game Developer

"My math degree gave me lots of career choices. One of the many options I had was to work in computer games. It might not seem like an obvious choice for a mathematician, but actually there's a lot of math involved in game graphics.

"There isn't really a typical working day in the game business. Every game has a development cycle, so the work depends on what part of the cycle we're in. It starts with the ideas or concept stage when we think up a new game. When we think we've got something good, we produce a demo to show the company what it's all about. If we get the green light to go ahead, the hard work to produce the game begins. It takes a dozen or so people about 18 months to produce a prototype. Then we hand it over to the testers, so they can find all the bugs and glitches.

"You can't rest on your laurels. Processors get faster every year and memory gets cheaper, so you can process more information faster and produce better games. The business is very competitive, because every company is trying to get ahead of all the others. We have to keep improving our games all the time, making them work faster and look better or we'll get left behind. That's what makes it worth coming to work every day."

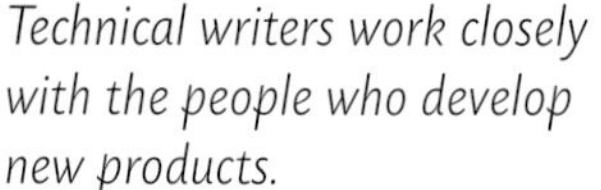

Technical writers work closely with the people who develop new products.

TECHNICAL WRITER

The software and hardware developed for ICT applications needs instruction manuals, brochures, and presentations so that other people can understand them and learn how to operate them. The people who write this material are technical writers. They're also called technical authors and document specialists. Technical writers find out what the users of a new system or software package need to know by talking to the people who developed it. Their job is then to write the information down clearly for the users. The users might be members of the public, nontechnical managers, or highly qualified engineers. The technical writer has to choose the level of language and technical content carefully to suit the users who will read it.

INSIDE KNOWLEDGE

A large part of a technical writer's work involves gathering information from system designers and software developers, so good people skills are an asset. It also helps if technical writers know something of the technology they are writing about. Once all the necessary information has been collected, the rest of the work involves writing the document in a logical and easy-to-understand form, so technical writers need excellent writing skills. If photographs, graphics, or drawings are necessary, the technical writer organizes the production of these too.

TO BECOME A TECHNICAL WRITER, YOU WILL NEED

- ***the ability to research and analyze information***
- ***a methodical approach to work***
- ***accuracy and attention to detail***
- ***excellent writing skills***

MAIN TASKS:
TECHNICAL WRITER

- *gathering and analyzing information*
- *writing new documents*
- *updating existing documents*

THE FINAL DRAFT

The text and illustrations are combined to make the first rough draft of the document. The project staff review the first draft and they will probably suggest changes. The document might go back and forth, being altered and refined by the technical writer several times before it reaches the final draft that everyone is happy with. The technical writer may have the final document printed on paper or prepared as an on-screen document stored on a CD or DVD, or available online as a download. In some companies, technical writers may also get involved in writing publicity brochures and developing web content.

SKILLED WRITERS

Technical writing requires people who are not only skilled writers, but who are also able to keep pace with fast-developing technology. Degrees in communications, journalism, or technical writing are attractive to employers, but so are scientific, technical, and engineering degrees. A knowledge of popular word processing, graphics, and desktop publishing packages is desirable, because technical writers use them to prepare documents. With some experience, technical writers can set up their own business. Freelance (self-employed) technical writers find their own work, set their own working hours and negotiate payment rates with employers. They may split their time between working at home and working at a client's offices.

Technical writers must keep the users of their documents in mind to ensure that the level and content of the document is appropriate.

Telecommunications technicians need a good head for heights, if their work includes dealing with antennas on top of towers and buildings.

Communications and Broadcasting

Gone are the days when a telephone was the only means of fast long-distance communication for most people. Today, homes and businesses use telephones, cell phones, the Internet, wireless devices, radio, satellite, and cable. Technicians and engineers install, maintain, and repair them.

TELECOMMUNICATIONS TECHNICIAN

Telecommunications technicians install, test, and repair communications equipment. They work on all types of telecommunications equipment, including telephone, cell phone, radio, satellite, and cable television systems. They have to understand IT systems, computer hardware, and software, because these are used extensively in telecommunications.

VARIETY OF WORK

Telecommunications technicians install telephone, broadband, and satellite equipment in homes and businesses and install telecommunications networks in large organizations. They lay underground communications cabling and repair breaks in telephone exchanges. And they install communications antennas on buildings and towers. They may have to work from engineering drawings or complex circuit diagrams created by designers and engineers, so they need to be able to read and understand these technical plans. Some technicians are based at their employer's offices. Others are field workers—they travel to customers' businesses or homes to install and repair equipment.

A FAST RESPONSE

A telecommunications technician's work is often a mixture of maintenance and repairing malfunctions. Maintenance involves testing equipment, looking for problems that might cause breakdowns, and replacing parts before equipment breaks down. Maintenance can be planned ahead of time, but malfunctions can happen at any time. Once a problem is reported, a technician has to respond without delay. Technicians may have to spend some time on-call. Being on-call means standing by, available for work, in case a problem is reported.

Telecommunications technicians carry out essential connection and maintenance work on telephone systems.

TO BECOME A TELECOMMUNICATIONS TECHNICIAN, YOU WILL NEED

- *manual dexterity*
- *computer skills*
- *numerical skills*
- *technical knowledge*
- *patience*

FINDING A JOB

A degree is not essential. Telecommunications technicians often start through an apprenticeship or an entry-level position with a company. Taking a relevant course at a technical school can give applicants an advantage. Training is on-the-job, usually while shadowing an experienced technician.

COMMUNICATIONS ENGINEER

Communications engineers, also known as telecommunications engineers, bridge the gap between technical staff and managers. They deal with the design, construction, and management of equipment that transmits, processes, or stores information. The information could represent almost anything. It might be telephone calls, television shows, computer data, Internet data, text, or images. It could be transmitted by electrical cable, optical cable, or radio, and stored digitally. Some communications engineers concentrate on management activities. They plan and supervise projects. Others work as technical experts. They solve practical communications problems.

BUSINESS NEEDS

If people in business want to set up a videoconferencing system, they turn to communications engineers to design it for them. If they have lots of telephone calls, they might ask communications engineers to create a more streamlined and efficient telephone system. Good communication is vital to the armed forces. Banks depend on high-speed communications for their cash machines and credit card payment systems in stores. The oil and gas industries have to keep in touch with drilling platforms and exploration rigs in remote locations all over the world. Communications engineers work in these and many other businesses and industries where communication is vital.

TO BECOME A COMMUNICATIONS ENGINEER, YOU WILL NEED

- *technical knowledge*
- *good communication, presentation, and people skills*
- *the ability to work under pressure and meet deadlines*
- *project management skills*

DESIGN AND INSTALLATION

A project begins with meetings where engineers learn more about an organization's communications needs. Good personal skills are desirable for this work. The engineers carry out a site survey. They look at the buildings and offices where a communications system will be installed. They take measurements and spot potential problems. Then they design a communications system to suit the site and the user's needs. A completely new system isn't always necessary. Often, engineers can upgrade or extend an existing system. Then the engineers present the proposed system to management for their approval.

SPECIALIZED WORK

Communications engineers work in research, development, and manufacturing. Most of them work for telecommunications companies and companies that manufacture and install communications equipment. Some communications engineers specialize in a particular industry or sector. Marine communications engineers concentrate on the shipping industry. There are rail communications engineers too. Space communications engineers specialize in spacecraft communications.

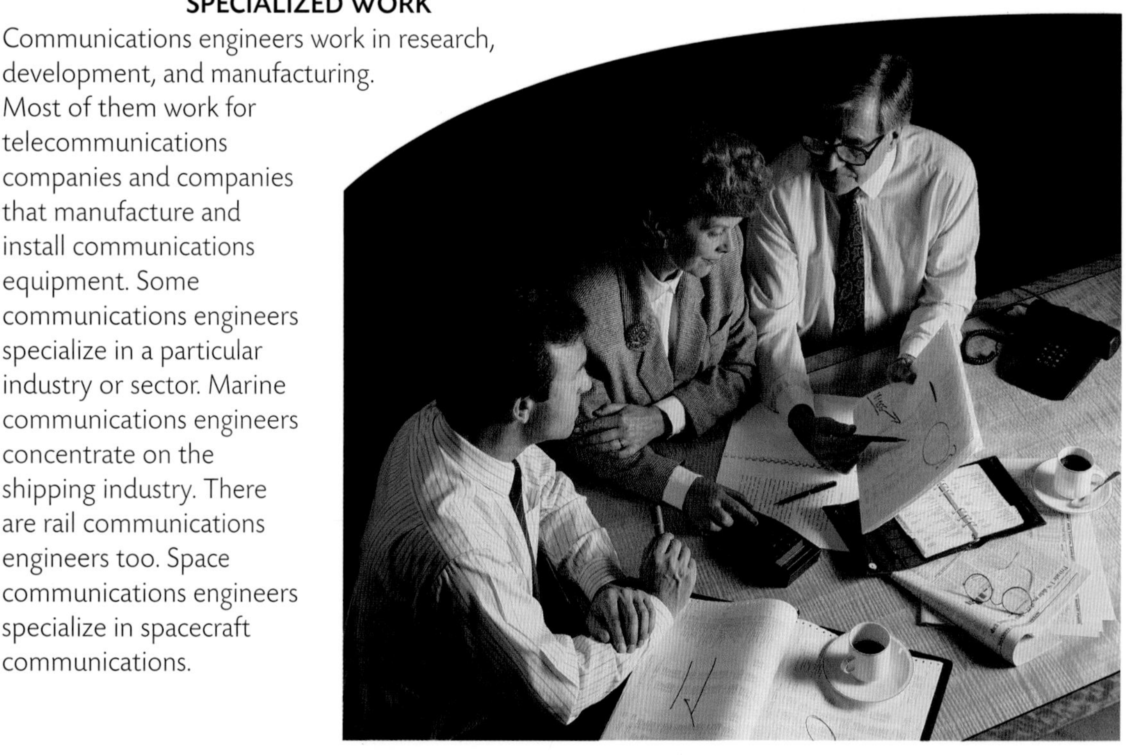

Communications engineers are briefed on a client's needs before designing the communications system needed.

Broadcast engineers do the behind-the-scenes technical work that makes radio and TV broadcasting possible.

BROADCAST ENGINEER

Broadcast engineers are electronics engineers and communications engineers who specialize in radio and television broadcasting. They are responsible for the engineering aspects of broadcasting. They make sure that radio and television programs are broadcast at the right times and ensure that the signal quality is up to the required standard. They work for broadcasting organizations, satellite and cable broadcasters, and independent production companies. Production companies provide broadcasting equipment, studios, and engineers for making radio and TV programs and commercials. Broadcast engineers have the opportunity to work in newer areas too, such as interactive programming, podcasting, and live feeds to the Web.

BROADCASTING EQUIPMENT

Broadcast engineers operate and maintain equipment used by broadcasters in studios, on outside broadcasts, and at transmitters. They set up audio and video links between studios, install all sorts of broadcasting equipment and software, and they service equipment and repair malfunctions. They set up microwave links between outside broadcast trucks and studios.

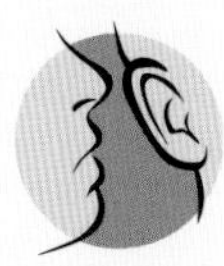

HANDY HINT

Communications engineers need good computer skills and software knowledge, because communications systems are monitored and controlled by computers.

STIFF COMPETITION

Some of the larger broadcasting organizations run internship programs for newly recruited broadcast engineers. There is stiff competition for these internships. Many of the successful applicants have already had some experience working with broadcasting equipment. They might have done some volunteer work for a small radio station or produced programs as part of a student project. Some may have worked at college radio stations.

STAYING UP-TO-DATE

Most broadcast engineers have a degree or similar technical certification in electrical engineering, electronics engineering, telecommunications engineering, computer engineering, or broadcast engineering. Alternatively, it may be possible to join a broadcasting organization as an assistant, researcher, or runner (messenger) and then get a place on an in-house training program. After that, further training is usually done on the job. Qualified broadcast engineers have to continue taking training courses to stay up-to-date with new equipment and technologies. Shift work is often necessary to provide engineering coverage, especially as it's common now for radio and television stations to be on the air 20–24 hours a day.

TO BECOME A BROADCAST ENGINEER, YOU WILL NEED

- *good practical skills*
- *knowledge of electronic and communications equipment*
- *good color vision and normal hearing*
- *good problem-solving skills*

WHERE WILL I BE?

Broadcast engineers with some experience become team leaders, supervising other engineers. Experienced engineers with good contacts may be able to go freelance and work for themselves. Alternatively, some engineers move into management positions.

Broadcast engineers work on outside broadcasts in all sorts of places.

Some broadcast engineers specialize in operating studio equipment.

Gareth: Broadcast Engineer

"While I was doing my degree in applied physics, I did some volunteer work on the engineering side of a local hospital radio station. I helped to keep the equipment going. When I graduated, I managed to get a place on a technical training course run by a broadcasting company. There were far more applications than places, so I was lucky to get in. I'm sure my hospital radio work helped. At the end of the course, the company offered me a full-time engineering position.

"The first thing I do when I come in every morning is to check my emails in case there have been any problems overnight that I need to know about. The engineers have regular meetings to discuss day-to-day progress and plan future work. One day I might be out on an outside broadcast, which could be anywhere in the country, operating the satellite truck that beams signals back to the studio. On another day, I might be installing a new mixing desk in an editing suite. A lot of the equipment is computer controlled, so some of the upgrade work that used to involve replacing equipment is now done by upgrading software.

"The work is interesting and varied. The hours are a bit irregular, because we all have to take our turn working late night or early morning shifts, weekends too—but I don't mind."

Working on the Web

The World Wide Web is an essential part of business and entertainment. Online shopping enables businesses, big and small, to reach customers all over the world. Millions of new web pages are created every year and they all have to be designed.

WEB DESIGNER

Web designers create the pages that make up a web site. They decide how the pages will look. They choose the colors, the layout, and the positions of text, images, buttons, and other links. They ensure that the pages are not too cluttered and make them easy to navigate through.

CREATING WEB SITES

Most web designers work for web design companies or the IT departments of large companies and organizations of all types. About a quarter of all web designers are freelance. Many of them work at home. Formal training is not essential to become a web designer. Many web designers are self-taught. However, training in graphic design or web design, or a basic technical certification is an advantage, as is a degree in a relevant fields such as computer science, web design, or multimedia.

WEB DEVELOPER

Web developers are often confused with web designers, but the two are different. Web developers are software engineers or programmers who write the software that makes a web site work.

TO BECOME A WEB DESIGNER, YOU WILL NEED

- *a flair for design*
- *good problem-solving skills*
- *creativity*
- *good computer skills*

MAIN TASKS: WEB DESIGNER

- *meeting clients to discuss their needs*
- *designing web sites for clients*
- *revising designs according to client wishes*
- *testing the final versions of web sites*
- *uploading web sites to servers*

Web designers use word processing, image handling, and design software to design web sites.

Web designers work alone at home or with other web designers in offices.

Sophie: Web Designer

"After graduating with a multimedia degree, I worked as a junior web designer for an IT consulting firm for about 18 months. It was great training. But when the company was taken over, I was laid off and decided to try going freelance. It was a big gamble. Setting up my own business was daunting, but I spoke to an accountant who guided me through it. The first few customers were difficult to find, but then word of mouth started bringing in more work.

"Working at home means that I can decide when I work, and I don't have to waste time commuting to and from an office somewhere else. I can work long hours when I need to and enjoy extra time off when I have less work.

"You have to enjoy working on your own and be able to motivate yourself to get work done on time. There are a couple of disadvantages to working for yourself. You have to buy all your own equipment and software, and you don't get any paid holidays. But I really enjoy going to work every day—all the way to the spare room!"

WHERE WILL I BE?

Web developers can move into management, perhaps becoming web content managers in charge of an organization's web site business development, or they can set up their own businesses providing web design and development services to clients.

Systems and Networks

Computers and communications equipment are connected together to form bigger systems and networks. Some IT professionals specialize in planning, building, and maintaining these systems and networks.

SYSTEMS ANALYST

Systems analysts advise businesses on their IT systems. They may also be known as business analysts or systems designers. They recommend the best system to make a business more efficient and productive. An IT system that might be suitable for a business one year may be inadequate a few years later because the business has grown or changed, and newer, better IT systems are available.

Systems analysts hold meetings with clients to find out what sort of IT system they need.

MAIN TASKS: SYSTEMS ANALYST

- *identifying a business's IT needs*
- *planning new IT systems*
- *testing IT systems*
- *training staff on new IT systems*

GATHERING INFORMATION

There are two parts to a systems analyst's work—analysis and design. First, a systems analyst learns more about the business, what it does, how it works, and what the people running it want a new or upgraded IT system to do. The analyst does this by talking to the staff and asking lots of questions about what they do and how they do it. Then he or she analyzes the business's IT needs. Second comes the work that involves designing the system, or an upgrade to the existing system, to do what is required. Analysts have to project costs, to keep them within the client's budget and plan the installation work. They attend regular meetings to discuss the questions and problems that inevitably arise during the design process. Once a team of technicians and engineers have installed the system, systems analysts supervise tests to ensure that it performs as it should.

JACK-OF-ALL-TRADES

A systems analyst is a jack-of-all-trades as well as a good problem-solver. He or she understands the pros and cons of various programming languages, operating systems, computer hardware, and communications equipment. To keep up-to-date with all the products analysts build systems from, they attend meetings with their hardware and software suppliers, and they work closely with programmers, designers, and engineers.

ONE OR MANY

Systems analysts are employed in all sorts of industries, commercial businesses, and government bodies. They may work for the organization they are advising or they may work for an IT consulting firm that provides professional advice to other companies. They work in an office environment, but they don't spend all their time sitting in front of a computer. They make visits to clients to gather information and supervise work. In small-to medium-sized organizations, one systems analyst may do everything—analyzing the business, designing a new or improved IT system, and managing the installation work. In large organizations, projects are usually too big for one analyst to handle, so a team of systems analysts works together on each project.

WHERE WILL I BE?

Systems analysts are promoted to senior analyst positions, or they specialize in particular types of business such as financial, transportation, retail, or manufacturing. Their technical expertise also lets them move into other areas such as marketing.

A systems analyst tours the business to see firsthand how it works and what the staff do.

Richard: Systems Analyst

"I particularly enjoy the trouble-shooting and problem-solving part of being a systems analyst. When clients come to me, they have a problem and it's my job to find a solution.

"I interview clients to find out what their requirements are. I talk to the people who will use the system to find out what they do. You have to feel comfortable talking to everyone from the receptionist to the CEO. I need to know what the system will have to do, who will use it, how many people will use it, how much information it has to handle and so on. I analyze the business operations too. It's like assembling all the pieces of a jigsaw puzzle. Then I draw up a proposal for an IT system that will deliver what the client needs. I work with programmers, hardware engineers, and other colleagues, and I draw on their expertise in these specialist areas.

"Clients sometimes have quite fixed ideas about what they want or what they think their business needs, but actually it may not be the best solution for them. You need good people skills to persuade them to accept a proposal for something different. You have to be a bit of a salesman as well as a technical professional. And you need to be able to be able to talk about IT systems in plain English instead of "geek-speak" when necessary."

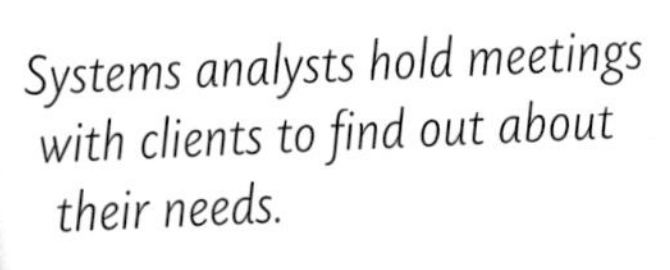

Systems analysts hold meetings with clients to find out about their needs.

A large organization's server room is full of computer equipment, which is installed under the supervision of systems analysts. They set up firewalls and other security software.

NETWORK ENGINEER

Network engineers build and maintain the networks that connect computers to each other and to the Internet. They deal with the "plumbing" that connects computers. In some cases, network engineers design networks too, but the biggest and most complex computer networks are designed by more senior IT professionals called network architects. Network engineers deal with computer networks called LANs and WANs. A LAN (Local Area Network) extends over one building or a group of buildings. A WAN (Wide Area Network) is bigger, perhaps extending over a whole country or beyond.

HIGH-SPEED COMMUNICATION

Computer networks enable people and machines to communicate with one another electronically, share files, and access the Internet. Even quite small companies have several computers linked to servers in a network. Larger businesses have networks that may spread across a city or country, or even beyond to other countries. As well as Internet access, they may also have their own intranets. These are private web sites or networks that use the same software as the Internet but cannot be accessed by the public. Network engineers are responsible for making sure that these networks work efficiently. They need good technical problem-solving skills and a wide knowledge of network equipment such as servers, routers, modems, hubs, and switches. They also need good personal communication skills, because they work within teams of other IT professionals and deal with nontechnical staff.

Network engineers maintain and upgrade complex communications systems.

MONITORING PERFORMANCE

A network engineer's day-to-day work involves monitoring the performance of computer networks and solving technical problems. Network engineers spend a lot of time drawing networks on white boards to visualize problems and work out solutions. When network engineers have to make changes to a network, they often have to do it outside normal office hours so that they don't interrupt other workers. Businesses that are too small to employ their own full-time network engineers use IT consulting firms that provide network services. Network engineers working for these companies look after several clients each. They make regular visits to each client to carry out routine checks and maintenance, and deal with any network issues brought up by the client's own IT support team.

MAIN TASKS: NETWORK ENGINEER

- *designing and building computer networks*
- *monitoring network performance*
- *monitoring network usage*
- *dealing with breakdowns*
- *planning upgrades*

The digital communications networks that link the computers in large organizations like libraries are designed by network engineers.

WHERE WILL I BE?
Experienced network engineers can move on to more senior IT design, engineering, or management positions. They can also specialize in one particular area of network engineering such as network security.

INFORMATION SCIENTIST

The "I" in IT and ICT stands for "Information." Some organizations hold so much information that they employ people to look after it and make use of it. These people are called information scientists or computer scientists. People who work mainly with computer databases, storing information such as customer records, are called database administrators. Information scientists work for government departments, educational establishments, libraries, large manufacturers, pharmaceutical companies, insurance companies, charities, and large professional practices such as law firms. If they work for an organization like a law firm or pharmaceutical company, they usually have some knowledge of the subject matter the organization deals with.

VARIED WORK

Information scientists deal with the storage, retrieval, and analysis of information from books, magazines, and digital storage systems. They carry out research, answer requests for information, and make sure the information stored in a system is kept up-to-date. So much information is stored or indexed electronically today that information scientists spend a lot of time sitting at computer screens as well as dealing with printed books, magazines, and journals. An important part of their work is to make sure that digital information storage systems operate within the laws and regulations that cover the sort of information that can be stored.

WRITING REPORTS

Information scientists don't just organize and store information. They also carry out research and write reports and briefing papers for the organizations they work for. The reports are used to study market trends, plan an organization's future strategies, and make comparisons with the activities and performance of competitors. This work requires patience, accuracy, and good writing skills.

GRADUATING TO INFORMATION

People sometimes find their way into this line of work after taking a temporary job in a library. Seeing the cataloging, indexing, and research work that goes on behind the scenes spurs them on to seek formal training and a career. The majority of information scientists have advanced degrees. They may have a degree in information science, information management, or a similar subject. Alternatively, they may have a completely unrelated undergraduate degree, which they follow up with a master's degree in information science.

TO BECOME AN INFORMATION SCIENTIST, YOU WILL NEED

- *computer skills*
- *writing skills*
- *an inquiring mind*
- *good people skills*
- *an understanding of the needs of information users*

MAIN TASKS: INFORMATION SCIENTIST

- *managing information resources*
- *cataloging and indexing information*
- *responding to requests for information*
- *writing and editing reports*
- *developing new information storage systems*
- *updating existing information storage systems*

Graduates may find it useful to gain experience working in a related field before starting out in a career as an information scientist.

An information scientist manages an organization's information resources, stored mainly in computers but also as books, magazines, and scholarly journals.

Natalie: Information Scientist

"I've been working as an information scientist for a medical research institute for the past three years. I spend about half of my time dealing with information requests. They arrive every day by phone, letter, and e-mail, mainly from researchers and journalists, but also occasionally from members of the public. When someone makes an inquiry, I use books, scientific and medical journals, our own databases, and the Internet to find the information they need. Tracking down the right information or the right publication for someone is very satisfying.

"The other half of my time is spent managing and developing our information resources. As new materials come in, they have to be indexed so that we know what we've got and, just as important, where to find it. It's no good holding lots of information if you can't find what you want. We maintain a database of references to articles published in magazines and journals. When publications come in, one of my jobs is to write short summaries of relevant articles and add them to the database. Then when someone sends in a request for information, we don't have to plow through hundreds of journals and printed indexes. We can search our database and get a list of all the relevant articles in seconds."

Further Information

BOOKS

Basta, Nicholas. *Careers in High Tech*. McGraw-Hill, 2007.

Careers in Focus: Computer & Video Game Design. Ferguson, 2009.

Careers in Focus: Telecommunications. Ferguson, 2009.

Croce, Nicholas. *Cool Careers Without College for People Who Love Video Games*. Rosen Pub. Group, 2007.

Discovering Careers for Your Future: Library and Information Science. Ferguson, 2008.

Heller, Steven, and David Womack. *Becoming a Digital Designer: A Guide to Careers in Web, Video, Broadcast, Game and Animation Design*. John Wiley & Sons, 2007.

Kirk, Amanda. *Field Guides to Finding a New Career: Information Technology*, Ferguson, 2009.

Kirk, Amanda. *Internet and Media*, Checkmark Books, 2009.

WEB SITES

http://www.bls.gov/oco/home.htm
The Bureau of Labor Statistics' Occupational Outlook Handbook lists the training and education needed, earnings, expected job prospects, descriptions, and working conditions of hundreds of jobs.

http://www.careercornerstone.org
The Sloan Career Cornerstone Center is an ever-expanding resource for anyone exploring careers in science, technology, engineering, mathematics, computing, and health care. Browse interviews with hundreds of professionals and download resources as PDFs, PowerPoints, and podcasts.

http://www.careervoyages.gov/infotech-main.cfm
The U.S. Department of Labor maintains current information on various fields, including information technology. Includes industry overviews, videos, sample career paths, and more.

http://www.comptia.org/home.aspx
CompTIA is the nonprofit trade association advancing the global interests of information technology (IT) professionals and companies including manufacturers, distributors, resellers, and educational institutions.

http://tcc.comptia.org/si_new.aspx
This tech career compass developed by CompTIA helps new workers assess their skills and find where they would best fit into the IT industry.

Glossary

algorithm step-by-step instructions for solving a problem

applications programmer person who writes a computer program to perform a particular task such as word processing

apprenticeship a training course that involves learning from more experienced people while working for a company

broadcaster a company that transmits radio and/or television programs to the general public

bug mistake in a computer program

cabling the cables or wires that connect computer and communications equipment together

certification a widely recognized qualification showing that an engineer has reached a certain level of knowledge or competence

cinematographer someone skilled in movie photography

client a customer, someone who receives services from an ICT professional

consultant an expert who gives professional advice to clients

database a searchable collection of data or records stored digitally

diagnostic software computer programs that monitor a computer system to find out if it is working correctly and, if not, to identify the problem

documentation manuals that describe how computer hardware or software works

embedded computer a computer that is built into another device such as a car or cell phone

engineer a person who uses scientific knowledge to solve problems.

entry level the first or lowest level of employment

forensic related to scientific techniques used in connection with legal proceedings. Forensic computer engineers examine computers seized during crime investigations for evidence that might be used in court.

freelance working independently for a number of employers

hardware the nuts, bolts, electronic circuits, and other equipment that make up a computer

internship a temporary position with a company, often taken while a student, to learn skills for a specific job

media platforms standard ways of distributing text, images and other information, including interactive television and radio

microwave link short frequency radio waves used to transmit audio or video for TV and radio broadcasts

people skills the abilities that make someone good at dealing with people, including the ability to explain things clearly

processor the central processing unit of a computer

prototype a trial version of a product, used for testing before the product is manufactured

Quality Assurance tester Someone who tests equipment or software to find bugs.

server a computer that supplies programs, files, and data to computers connected to it

shift work work periods outside normal office hours

software computer programs. The sets of instructions that control computers.

software architect designer of computer programs

supercomputer one of the fastest types of computers in the world

systems analyst someone who studies a business and recommends the best computer systems and software for its needs

technical support help for people who experience problems when using computer and communications equipment and software

technical writer a writer who produces documentation for hardware, software, and systems

technician a person with a working practical knowledge of systems, hardware, and software, enabling them to carry out installation, maintenance, and repair work

telecommunications the science and technology of communicating over long distances electronically

upgrade software and/or hardware that improves on what was used before

Index